Chemical Engineering & the Law Forum 2017

Core Programming Area at the 2017 AIChE Annual Meeting

Minneapolis, Minnesota, USA
29 October – 3 November 2017

ISBN: 978-1-5108-5810-7

Printed from e-media with permission by:

Curran Associates, Inc.
57 Morehouse Lane
Red Hook, NY 12571

Some format issues inherent in the e-media version may also appear in this print version.

Printed by Curran Associates, Inc. (2018)

AIChE
120 Wall Street, FL 23
New York, NY 10005-4020

Phone: (800) 242-4363
Fax: (203) 775-5177

www.aiche.org

Additional copies of this publication are available from:

Curran Associates, Inc.
57 Morehouse Lane
Red Hook, NY 12571 USA
Phone: 845-758-0400
Fax: 845-758-2633
Email: curran@proceedings.com
Web: www.proceedings.com

TABLE OF CONTENTS

Some abstracts in this book refer to images that were not made available by the publisher

509054 Solve this! Fundamental Approach to Problem Solving in Industrial Processes I (Invited Talks)

Abstract:

Solving complex plant problems is an everyday exercise that affects all projects and budgets, and it is very rarely introduced to students at any level. In fact, by the time young engineers are few years into working in industry, most have lost the engineering skills that they developed in school. We hope to encourage future engineers to retain those skills and learn the importance of plant problem-solving.

WENDEROTH

ONE FOCUS | PROSECUTION Patents | Trademarks

Proactively Preparing Experimental Data in Chemical Patent Applications

Peter M. Jay
pjay@wenderoth.com

Effective Patent Applications

- "There are few, if any, legal documents more difficult to craft, more fraught with pitfalls than patent applications . . . ."
  - *Energizer Holdings, Inc. v. International Trade Commission*, 2008 WL 1791980, *10 (Fed. Cir. April 21, 2008)(unpublished) (J. Newman) (dissenting).
- What are some of the pitfalls, and how can Applicants proactively avoid them?

WENDEROTH

Hypothetical Scenario

- After years of research, you discover an amazing improvement on January 1, 2017.
- You prepare an invention disclosure and contact your patent attorney about one month later.
- The patent attorney drafts a patent application and files the application at the USPTO on March 1, 2017.
- The USPTO issues a substantive Office Action on September 1, 2018, and experimental data is needed to respond.

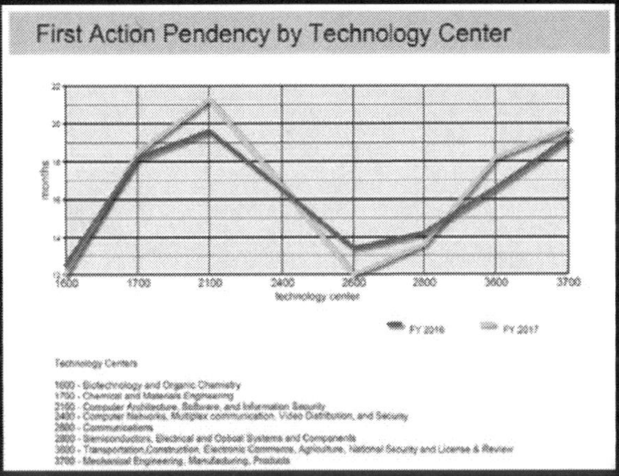

Source: https://www.uspto.gov/corda/dashboards/patents/kpis/kpiTCFirstActionPendency.kpixml

WENDEROTH

What Happened Between January 1, 2017, and September 1, 2018?

- In the *20 months* from January 1, 2017, to September 1, 2018, many changes could occur:
  - Grants expire/budgets are refocused
  - Research assistants graduate/co-inventors leave the company
  - Laboratory equipment is repurposed
  - The mind begins to forget details
  - Etc.
- In reality, producing experimental data can be difficult *at any time*.
- However, obtaining new experimental data when the inventors have moved onto other projects is often the most difficult.

WENDEROTH

Filing Patent Applications Worldwide

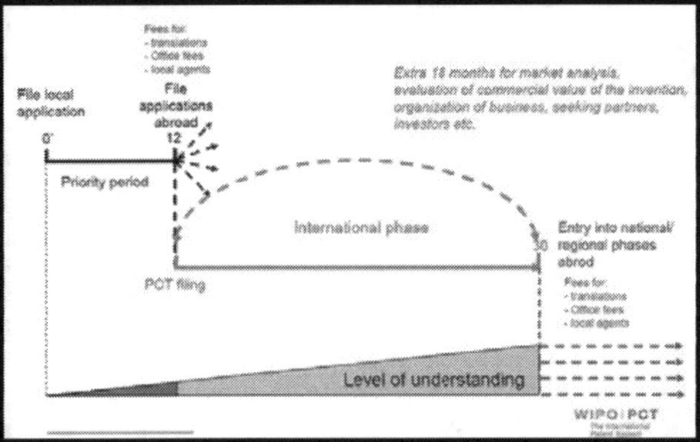

Source: Peter Waring, WIPO: Overview of the PCT System (June 26, 2013)

Table 4.3: STATISTICS ON PROCEDURES

Definitions of the various terms are given in Annex 2.

Progress in the procedure Rates in percentage	Year	EPO	JPO	SIPO	KIPO	USPTO
Examination	2014	93.1	67.9	682,158	80.8	100
	2015	93.8	69.4	809,661	82.5	100
Grant	2014	47.6	69.3	233,228	68.6	70.9
	2015	48.0	71.5	359,316	65.0	70.6
Opposition	2014	4.7	-	-	-	n.a.
	2015	4.4	-	-	-	n.a.
Appeal on examination	2014	22.1	26,174	-	11.6	3.7
	2015	20.0	22,263	-	11.5	2.7
Pendency in the procedure						
Awaiting request for examination	2014	139,038	701,836	n.a.	286,270	-
	2015	34,438	694,255	n.a	285,816	-
Pending examination	2014	396,049	186,830	n.a.	171,178	610,227
	2015	431,632	193,390	n.a.	161,770	565,811
Pendency first action (months)	2014	9.1	9.6	12.5	11.0	18.1
	2015	9.4	9.5	12.8	10.0	18.4
Pendency final action (months)	2014	22.8	15.2	21.8	16.7	27.0
	2015	26.4	15.0	21.8	14.1	26.3
Pendency invalidation (months)	2014	-	-	6.4	-	-
	2015	-	-	5.4	-	-

- = not applicable n.a. = not available

Source: IP5 Statistics Report 2015

- Filing patent applications worldwide can result in *even more delays*. A delay of about *40 months* is not unusual.
- Some countries (e.g., Japan) limit the effect of post-filing data submission.

WENDEROTH

Hypothetical Scenario

- After years of research, you discover an amazing improvement on January 1, 2017.
- You prepare an invention disclosure and contact your patent attorney about one month later.
- The patent attorney drafts a patent application and files the application at the USPTO on March 1, 2017.
- The USPTO issues a substantive Office Action on September 1, 2018, and experimental data is needed to respond.

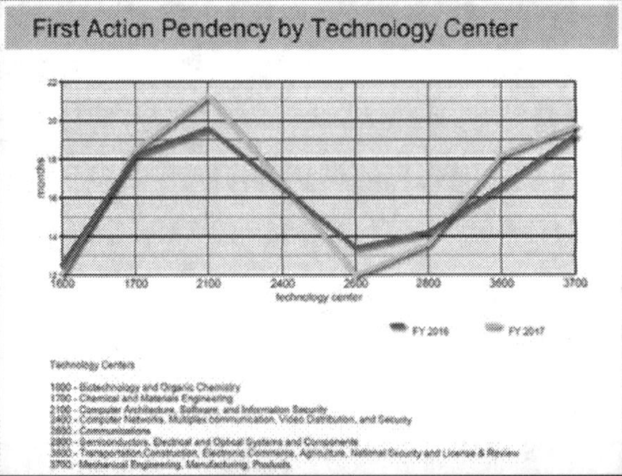

Source: https://www.uspto.gov/corda/dashboards/patents/kpis/kpiTCFirstActionPendency.kpixml

WENDEROTH

The Office Action

- Office Actions explain why, in a Patent Examiner's opinion, a patent application is not allowed to issue as a patent.
- The reasons provided are divided into rejections based upon either *form* or *substance*.
- Common issues in chemical applications:
  - Form
    - Definiteness
    - Enablement
  - Substance
    - Inherency
    - Unexpected Results

WENDEROTH

The Definiteness Requirement

- "The specification shall conclude with one or more claims particularly pointing out and distinctly claiming the subject matter which the inventor or a joint inventor regards as the invention." 35 U.S.C. §112(b).
  - The primary purpose: to ensure that the scope of the claims is clear so that infringement can be determined.
  - A secondary purpose: to provide a clear measure of what applicants regard as the invention.

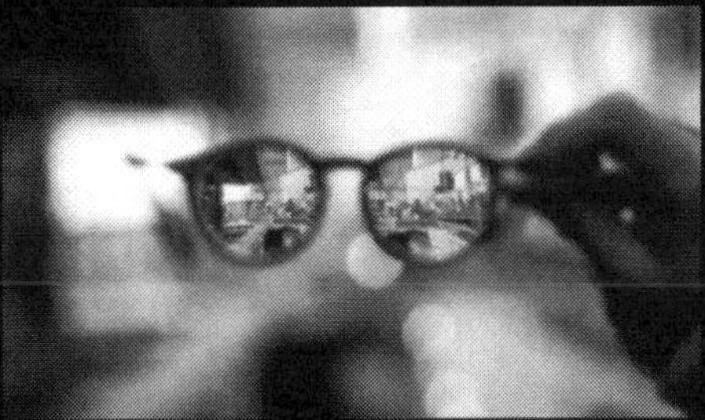

WENDEROTH

Clarity Matters – What is "molecular weight?"

What is claimed is:

1. A method of manufacturing copolymer-1, comprising reacting protected copolymer-1 with hydrobromic acid to form trifluoroacetyl copolymer-1, treating said trifluoroacetyl copolymer-1 with aqueous piperidine solution to form copolymer-1, and purifying said copolymer-1, to result in copolymer-1 having a molecular weight of about 5 to 9 kilodaltons.

Source: US Pat. No. 5,800,808

- What does the term "molecular weight" mean?
- "Molecular weight" can be ascertained by any of three possible measures: Mp (peak), Mn (number), and Mw (weight).
  - The claims do not indicate which measure to use.
  - The specification never defines molecular weight or even mentions Mp, Mw, or Mn.
  - And the term "average molecular weight" does not have a plain meaning to one of skill in the art.
- Claim 1 was invalid because the patentee has failed to inform with *reasonable certainty* those skilled in the art about the scope of the invention.

Source: *Teva Pharm. USA, Inc. v. Sandoz, Inc.*, 135 S. Ct. 831 (2015).

WENDEROTH

Clarity Matters – How are properties measured?

The slope of strain hardening coefficient (SHC) is calculated according to the following equation:

$$SHC = (\textit{slope of strain hardening}) * (I_2)^{0.25}$$

where I_2=melt index in grams/10 minutes.

Source: US Pat. No. 5,847,053

- Even though "strain hardening" and "melt index" are well-known parameters, the max *slope* could be determined via four different method.
  - Each method produced a different result.
  - How do you choose the correct method?
- Claim 1 found to be invalid because the patent did not discuss the four methods or provided any guidance about which method should be used.

Source: *Dow Chemical Co. v. Nova Chemicals Corp.*, 803 F.3d 620 (Fed. Cir. 2015

We claim:

1. A film made from an ethylene polymer composition, wherein the composition comprises (A) from about 10 percent (by weight of the total composition) to about 95 percent (by weight of the total composition) of at least one homogeneously branched linear ethylene/α-olefin interpolymer having:

(i) a density from about 0.89 grams/cubic centimeter (g/cm^3) to about 0.935 g/cm^3,

(ii) a molecular weight distribution (M_w/M_n) from about 1.8 to about 2.8,

(iii) a melt index (I_2) from about 0.001 grams/10 minutes (g/10 min) to about 10 g/10 min,

(iv) no high density fraction,

(v) a single melting peak as measured using differential scanning calorimetry, and

(vi) a slope of strain hardening coefficient greater than or equal to 1.3; and

(B) from about 5 percent (by weight of the total composition) to about 90 percent (by weight of the total composition) of at least one heterogeneously branched linear ethylene polymer having a density from about 0.93 g/cm^3 to about 0.965 g/cm^3.

Source: US Pat. No. 5,847,053

WENDEROTH

The Disclosure Must Be Clear

- Easy steps to avoid issues with clarity:
  - Keep detailed records while conducting experiments, including the methodology for each measurement.
    - For example, if a measurement was conducted according to ASTM or JIS standard, then include the standard.
  - Be careful when crafting new parameters.
    - If possible, try to use well-established parameters rather than novel parameters.
  - Ensure that your patent attorney has sufficient time to carefully review the record.

WENDEROTH

The Enablement Requirement – Quid Pro Quo

- "The specification shall contain a written description of the invention, and of the manner and process of making and using it, in such full, clear, concise, and exact terms as to enable any person skilled in the art to which it pertains, or with which it is most nearly connected, to make and use the same . . . ." 35 U.S.C. §112(a).
- Is the experimentation needed to practice the invention undue or unreasonable? *See* MPEP § 2164.01.
- To make this determination, the USPTO considers the *Wands* factors:
  - (A) The breadth of the claims;
  - (B) The nature of the invention;
  - (C) The state of the prior art;
  - (D) The level of one of ordinary skill;
  - (E) The level of predictability in the art;
  - (F) The amount of direction provided by the inventor;
  - (G) *The existence of working examples*; and
  - (H) *The quantity of experimentation needed to make or use the invention based on the content of the disclosure.*

In re Wands, 858 F.2d 731, 737 (Fed. Cir. 1988)

WENDEROTH

Using Examples to Enable the Invention

- Compliance with the enablement requirement *does not* turn on whether an example is disclosed. MPEP § 2164
  - Both "working examples" and "prophetic examples" can be used.
- Because chemistry is considered an "unpredictable art," working examples are often helpful in the chemical arts to enable broad claims.
  - Working examples can limit the amount of experimentation needed to practice the entire scope of the invention.
- In almost all chemical applications, at least one working example should be included. The working example should, if possible, be directed to a commercial embodiment to ensure that the commercial embodiment is enabled.
- More examples are better in most circumstances. More variety is better in most circumstances.

WENDEROTH

Early Experimental Evidence

- MPEP § 716 provides: "Examiners must consider comparative data in the specification which is intended to illustrate the claimed invention in reaching a conclusion with regard to the obviousness of the claims.
- MPEP § 2145 provides that rebuttal evidence and arguments can be presented in the specification or by way of an affidavit or declaration, and that it is error not to consider evidence presented in the specification.

By anticipating substantive issues, experimental evidence can be prepared and submitted *as part of the application*!

WENDEROTH

Duty of Disclosure, Candor, and Good Faith

- Don't hide the ball!
- Must disclose all information that refutes, or is inconsistent with, a position the applicant takes in:
  - (i) Opposing an argument of unpatentability relied on by the Office, or
  - (ii) Asserting an argument of patentability.
- Lovenox® patents invalidated for providing misleading information.
  - Compared half-life with a known drug, but failed to disclose that different doses were used.

WENDEROTH

Being Prepared for Inherency

- The claiming of a new use, new function or unknown property which is inherently present in the prior art does not necessarily make the claim patentable. MPEP § 2112

- Inherency is fundamentally based upon logic.
- If two things are the same, then they must necessarily have the same properties.
- The USPTO provides that an inherent feature need not be recognized by one of ordinary skill in the art at the time of the invention.
- However, the inherent feature must *necessarily* be present in the prior art—not a mere probability or possibility.

WENDEROTH

Inherency Example

- A patent application contains the following claim:

  1. A composition comprising A, B, and C, wherein the composition has a compression strength of 100 MPa to 250 MPa.

- The prior art discloses compositions comprising A, B, and C, but fails to discuss the compression strength.
- Does the prior art inherently disclose the compression strength?
  - It depends!?!?

WENDEROTH

Inherency: Asking Why?

- Asking "why?" can help lead the inherency discussion.
- For instance, the prior inherency example could be aided by asking:
  - Why did the combination of A, B, and C achieve the compression strength?
    - Did the reaction conditions matter?
    - Did the order of adding the components matter?
  - Why wouldn't the compression strength necessarily be achieved by *all* combinations of A, B, and C?
- The answers to these "why?" questions can focus the discussion in the application and focus early experimental considerations.

WENDEROTH

Proactive Unexpected Results

- Applicants can present evidence to rebut a finding of obviousness. MPEP § 2145
- Evidence of unexpected results is the most prevalent evidence of non-obviousness, and is particularly common in the "unpredictable arts," such as chemistry.
- The evidence should provide a comparison to the closest prior art. However, it is almost impossible to predict with accuracy the exact document or combination of documents that will be considered the closest prior art.
- Use the patent application to lay the framework for unexpected results with a wide variety of data, and *supplement* the framework with additional declarations.

WENDEROTH

Commensurate in Scope

- Whether the unexpected results are the result of unexpectedly improved results or a property not taught by the prior art, the "objective evidence of nonobviousness must be commensurate in scope with the claims which the evidence is offered to support."
- For instance, consider the following claim:

  A composition comprising:
  10-15% by weight of component A; and
  20-40% by weight of component B.

- Does the unexpected result exist when the composition contains
  - 10% of component A and 40% of component B?
  - 14% of component A and 32% of component B?
  - Etc.
- If an unexpected result will need to be argued, then *vary the experimental data* so that the data can be commensurate in scope with the claims.

WENDEROTH

Criticality of Ranges

- To establish unexpected results over a claimed range, applicants should compare a sufficient number of tests both inside *and* outside the claimed range to show the criticality of the claimed range.
- Lilienthal's Rule: If you want to fit a straight-line to your data, be certain to collect only two data points. A straight line can always be made to fit through two data points.
- Proving criticality will almost always require *more than two* data points.
- If we are trying to establish that the 15% by weight is critical in the prior exemplary claim, we will likely need at least three data points.
  - For instance, providing data at 11%, 14%, and 16% could be helpful.
  - If criticality will be an issue, prepare as many data points as possible to provide flexibility with claim amendments.

WENDEROTH

Good Recordkeeping

- The US is a first-to-file jurisdiction, but laboratory notebooks are still important.
- A laboratory notebooks can:
  - fill the gaps in experimental data and refresh the mind;
  - help establish correct inventorship;
  - establish the right to a trade secret;
  - preserve the prior use defense;
  - be used as evidence in litigation; and
  - support derivation proceedings.
- If your employer has a recordkeeping policy, *follow the policy*.

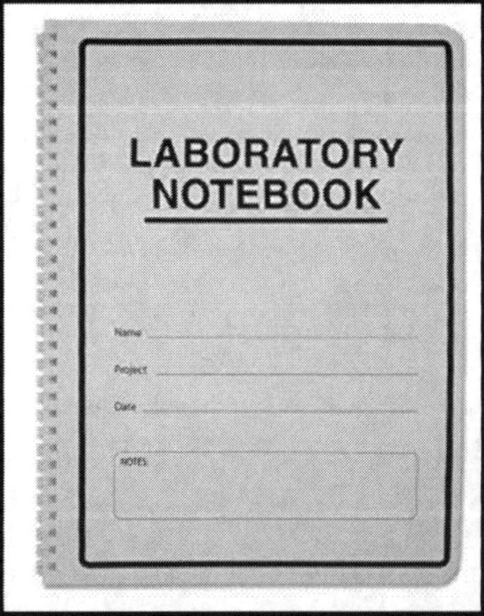

WENDEROTH

Tips for Preparing Experimental Data for Chemical Patent Applications

- Tip #1: Conduct as many experiments as possible with as much variety as possible.
- Tip #2: Keep detailed records of experimentation.
- Tip #3: Ask "why?"
- Tip #4: Contact your patent attorney early.

WENDEROTH

Questions?

pjay@wenderoth.com

WENDEROTH

AUTHOR INDEX